This document is in the public domain and may be freely copied

Disclaimer

Mention of any company or product does not constitute endorsement by the National Institute for Occupational Safety and Health (NIOSH). In addition, citations to Web sites external to NIOSH do not constitute NIOSH endorsement of the sponsoring organizations or their programs or products. Furthermore, NIOSH is not responsible for the content of these Web sites. All Web addresses referenced in this document were accessible as of the publication date.

Ordering Information

To receive documents or other information about occupational safety and health topics, contact NIOSH at

> Telephone: **1-800-CDC-INFO** (1-800-232-4636)
> TTY: 1-888-232-6348
> e-mail: cdcinfo@cdc.gov
>
> or visit the NIOSH Web site at **www.cdc.gov/niosh**.

For a monthly update on news at NIOSH, subscribe to NIOSH *eNews* by visiting **www.cdc.gov/niosh/eNews**.

DHHS (NIOSH) Publication No. 2009-135

May 2009

SAFER • HEALTHIER • PEOPLE™

Software Disclaimer of Liability Clause

The Determination of Sound Exposures (DOSES) software was developed by the National Institute for Occupational Safety and Health (NIOSH). It is provided "as is" without warranty of any kind, including express or implied warranties of merchantability or fitness for a particular purpose. By acceptance and use of this software, which is conveyed to the user without consideration by NIOSH, the user expressly waives any and all claims for damage and/or suits for personal injury or property damage resulting from any direct, indirect, incidental, special or consequential damages, or damages for loss of profits, revenue, data or property use, incurred by the user or any third party, whether in an action in contract or tort, arising from access to, or use of, this software in whole or in part.

No further development or upgrades for this software are planned. Any questions concerning this product can be directed to the NIOSH Pittsburgh Research Laboratory, Hearing Loss Prevention Branch (412-386-6560).

System Requirements

To run the DOSES program effectively, your computer should meet the following system requirements:

Operating system	Windows 2000/XP/Vista	
Hardware	**Minimum**	**Recommended**
Processor	Pentium 200 MHz or equivalent	Pentium III or greater or Celeron 600 MHz or equivalent
Disk space	80 MB	80 MB
Memory	64 MB of RAM	256 MB of RAM
Screen resolution	800 × 600	1024 × 768

Determination of Sound Exposures (DOSES):
Software Manual and Implementation Guide
By Gregory P. Cole, Ellsworth R. Spencer, and Eric R. Bauer, Ph.D.

CONTENTS

	Page
INTRODUCTION	1
COLLECTING TIME-MOTION DATA	2
Task Observations Versus Location Observations	2
Timeframe Recorded	2
Duration	2
Data Sheets	3
Instrumentation	4
INTERPRETATION OF RESULTS	4
Interventions	5
Intervention Examples	5
Intervention Success	5
USER INSTRUCTIONS	6
1. Start DOSES (Splash Screen)	6
2. OPEN Time Study	6
3. Time Study Properties - New Study Screen	7
4. Time Study Editor	10
5. Time Study Analysis	15
6. Frequently Asked Questions	18
REFERENCES	21
APPENDIX.—MSHA Equipment and Occupation Codes	22

ILLUSTRATIONS

1. Example of Time-Motion Study Sheet	3

TABLES

1. Dosimeter Settings	4

ACRONYMS AND ABBREVIATIONS USED IN THIS REPORT

AL	action level
DOSES	Determination of Sound Exposures
MSHA	Mine Safety and Health Administration
NEC	not elsewhere classified
NIOSH	National Institute for Occupational Safety and Health
PEL	permissible exposure level
REL	recommended exposure limit
SL	sound level
SLM	sound level meter

UNIT OF MEASURE ABBREVIATIONS USED IN THIS REPORT

dB	decibel
dB(A)	decibel, A-weighted
hr	hour
m	meter
MB	megabyte
MHz	megahertz

INTRODUCTION

The Determination of Sound Exposures (DOSES) software was developed by the National Institute for Occupational Safety and Health (NIOSH) specifically for use by mine management and safety personnel. DOSES simplifies the record-keeping and analysis associated with time-motion studies and worker noise exposures, making it easier to identify and solve noise problems.

The software relies on a time-motion study that profiles the worker's daily activities. At the same time, noise measurements are collected with a dosimeter or sound level meter (SLM). Observations about the worker's location and tasks or other activities are recorded along with times and durations so they can later be matched up with the noise data.

After the completion of a time-motion study of the worker's daily tasks and locations (possible noise sources), which measures the worker's noise exposures during the recorded events, the information is entered into DOSES. The program then displays information about the worker's accumulated noise dose over time.

When the sound level (SL) is constant over the entire work shift, the daily worker's noise dose (D), in percent, is given by:

$$D = 100 \, C/T$$

where   C = the total length of the workday, in hours,
and   T = the reference duration corresponding to the measured SL.

When the daily noise exposure consists of periods of different noise levels, the daily dose (D) is calculated according to the following formula:

$$D = 100 \times (C_1/T_1 + C_2/T_2 + \ldots + C_n/T_n)$$

where C_n = total time of exposure at a specified noise level,
and T_n = exposure time at which noise for this level becomes hazardous.

The software gives the user the option of assessing dose relative to the NIOSH recommended exposure limit (REL), the Mine Safety and Health Administration (MSHA) permissible exposure level (PEL), or the MSHA action level (AL). To obtain the NIOSH REL dose the software uses a 3-dB exchange rate, an 80-dB threshold level, and an 85-dB criteria level [NIOSH 1998]. To obtain the MSHA PEL dose the software uses a 5-dB exchange rate, a 90-dB threshold level, and a 90-dB criteria level. To obtain the MSHA AL dose the software uses a 5-dB exchange rate, an 80-dB threshold level, and a 90-dB criteria level [64 Fed. Reg.[1] 49548 (1999)]. The software uses the following formulas to calculate dose:

NIOSH REL dose = 0 for dB(A) < 80

                       = $(\text{Duration} / (8 / (2^{(L_{eq} - 85)/3}))) * 100$

MSHA PEL dose = 0 for dB(A) < 90

                       = $(\text{Duration} / (8 / (2^{(L_{eq} - 90)/5}))) * 100$

MSHA AL dose   = 0 for dB(A) < 80

                       = $(\text{Duration} / (8 / (2^{(L_{eq} - 90)/5}))) * 100$

where duration is in units of hours and L_{eq} is the SL taken from the SLM or dosimeter data.

---

[1]Federal Register. See Fed. Reg. in references.

The software generates a variety of interactive on-screen displays showing where, when, and how the worker's noise dose accumulated. It also can generate customizable printed reports. These outputs can be used to highlight the tasks, locations, and times that are associated with the greatest amount of the worker's noise exposure. Mine safety personnel can then use these reports to make decisions about how to reduce or eliminate the factors that are creating an overexposure.

COLLECTING TIME-MOTION DATA

As with any computer program, the utility and outcome of DOSES depends greatly on the quality of information entered. Task observation, or time-motion study, is the basis for the program's input and must be complete and detailed. Gaps in observations and recorded durations will produce poor results, comparisons, and misguided and ineffective exposure reduction attempts. For effective time-motion studies, the following must be considered: task observations or location observations, timeframe recorded, and full- or partial-shift observations.

Task Observations versus Location Observations

Depending on the classification of the worker being observed, it may be better to use one method of observation over another. Most of the time it is recommended and logical to use task observations. This is when the duration of each task that the worker performs is recorded. Observing a machine operator, for instance, would logically involve task observations because these operators tend to remain with the machine, positioned at a relatively fixed location, and perform a uniform and repeated series of tasks throughout their shift. For example, a continuous miner operator, even operating remotely, would stay at a relatively fixed position with respect to the continuous miner and complete tasks such as tramming, cutting, loading, cutting and loading, waiting on shuttle car, etc.

Location observations best suit time studies of workers who move about frequently during the shift, perform many different tasks, and spend time at many locations. Examples include utility workers in underground coal mines or plant workers at coal preparation plants. These workers are on the move during the entire shift but return to specific locations often during the shift. They may repeat similar tasks, but tend to repeat them at the same locations throughout the shift.

It is recommended that no matter which method of observation is used, detailed notes should be taken. When task observations are employed, the specific location of those tasks should be noted on the data sheets. Conversely, when location observations are used, the specific tasks being performed at each location must be noted. The end goal is to be able to get an accurate picture of each component of the employee's noise exposure so that crucial noise exposures can be managed and the difference in total dose (baseline versus control) can be ascertained.

Timeframe Recorded

A second important part of the data collection process is determining the timeframe for recording each task/location. At a minimum, times should be recorded to the minute. Even better is to note the time intervals to the second, i.e., in the format hr:min:sec. As experience is gained conducting observations for various occupations, the precision and detail of both the observations and recording times will be easier to determine based on length of each task or time at each location. Long-duration tasks/locations likely require a less stringent time recording (hr:min), while rapidly occurring tasks or changing locations would require a "tighter" recording time (hr:min:sec).

Duration

The duration of the observations must be considered. Full-shift observations take just that—a full shift—to complete, plus the time needed to enter the observation data and dosimeter files into DOSES. This is a substantial time commitment for the industrial hygienist or safety person. It can be difficult to free up an entire shift to complete the observations, but it is recommended that full-shift observations be done. Not only does this result in a complete description of what the worker does during a shift, it also gives the full-shift dose for compliance purposes.

For certain workers, partial-shift observation can still be useful when time does not permit full-shift observation. If a worker is known to perform the same tasks, the same number of times, and for the same duration each time during a shift, it may be possible to ascertain that worker's exposures with a partial-shift observation. Observing several complete cycles of each of the tasks performed may be sufficient.

Data Sheets

The abbreviated data collection sheet shown in Figure 1 is an example of the type of information that should be recorded when completing a time-motion study for use in the DOSES program. The bold items are standard to the sheet, while the nonbold items would be filled in by the observer. DOSES is designed for time-motion study data to be entered by start/stop time. The start and stop time of each work task and/or location and the related SL measurement are critical to the final analysis report. It is important to keep the start and stop times sequential without any gaps. DOSES uses the total time of a task or location to determine its contribution to the worker's daily dose. Leaving gaps between the time periods or entering the time period without a task or location will reduce the effectiveness of the DOSES program. It is also important to attempt to list all possible tasks or locations on the data sheets before beginning the study and to leave some blank columns for unanticipated tasks or locations. The program also requires entries for the date of the study, the employee's MSHA occupation code, and the MSHA equipment operated code (see appendix for full code lists). These data are needed to complete the DOSES analysis of the worker's work tasks and/or location contributions to the cumulative daily dose and for the mine's noise control database.

Data Sheet		Observations for:	Continuous Miner Operator			
Date: 11/11/2007	**Mine:** XYZ		**Section:** North 11		**Emp. First Name:** Sam	
Unique ID: 1234	**Manufacturer:** Smith		**Model:** XYZ-1		**Serial No.:** 12345	
Normal Shift: [Y]	**Shift Production:** 2,000 tons				**Dosimeter Tag:** 12345	
Observer: Jane Doe			**Other Info:** Pittsburgh Coal Seam 3.4 m			
		Task - Location Observations				**Sound Level Measurements**
Time Start	**Time Stop**	Mantrip	Downtime	Con. Miner Operator Face	Tramming Crosscut	
06:00:00	06:25:20	X				95 dB(A)
06:25:20	07:00:45		X			73 dB(A)
07:00:45	11:00:00			X		97 dB(A)
----	----	----	----	----	----	----
15:05:05	15:20:05				X	92 dB(A)
15:20:05	15:55:10		X			75 dB(A)
15:55:10	16:20:45	X				97 dB(A)

Figure 1.—Example of Time-Motion Study Sheet.

When using the time-motion study sheet, there is a need to be consistent, especially within a specific job classification. First, it is important to always label the worker occupations with the same names using the MSHA occupation codes (see appendix). Secondly, it is suggested that the exact same task names or location names be used. This allows for comparisons between time studies of the same employee and between employees with the same occupations.

Instrumentation

The instrumentation best suited for conducting exposure and SL analyses includes a person-wearable dosimeter for monitoring worker noise exposure and an SLM for measuring the SLs for each task or at each work location. The brand of dosimeter is not important. DOSES allows for downloading of dosimetry data from all of the dosimeters commonly used at mining operations because they have received MSHA intrinsic safety approval for use in underground coal mines. Most unlisted dosimeters can also be used as long as they provide time-stamped sound pressure level data in plain ASCII text files. Therefore, companies can choose their own SLM or dosimeter for measuring SLs.

Dosimeter settings for the NIOSH REL, MSHA PEL, and MSHA AL are shown in Table 1. For most mining applications, the MSHA PEL and AL are the dosimeter settings likely to be used, although DOSES does not require any particular setting as long as a time and sound pressure level are being saved to the dosimeter's internal memory. DOSES calculates the dose relative to three criteria—NIOSH REL, MSHA PEL, and AL—from the time and SLs downloaded from the dosimeter.

Table 1.—Dosimeter Settings

Designation	Parameters	Settings
NIOSH REL	Weighting	A
	Threshold level	80 dB
	Exchange rate	3 dB
	Criterion level	85 dB
	Response	Slow
	Upper limit	140 dB
MSHA PEL	Weighting	A
	Threshold level	90 dB
	Exchange rate	5 dB
	Criterion level	90 dB
	Response	Slow
	Upper limit	140 dB
MSHA AL	Weighting	A
	Threshold level	80 dB
	Exchange rate	5 dB
	Criterion level	90 dB
	Response	Slow
	Upper limit	130 dB

INTERPRETATION OF RESULTS

The primary outcome of DOSES is an understanding of a worker's noise exposure, what causes that exposure, or locations where the exposure is occurring. The charts, graphs, and tables generated by DOSES are all designed to display this important information. Specifically, the outputs summarize a worker's tasks/locations, time at each task/location, and dose accumulated at each task/location. This information is initially displayed as a compilation by tasks/locations, listed from highest accumulated dose task/location to lowest accumulated dose task/location. This information can be sorted by task/location, duration, or accumulated dose. Mine safety personnel should be looking for the tasks or locations that result in the most accumulated dose. Once this determination has been made, then the appropriate intervention strategies can be selected.

Interventions directed at the tasks or locations that result in the most dose have the best chance of significantly reducing a worker's noise exposure.

Interventions

Interventions fall into one of two types: engineering controls or administrative controls. Engineering noise controls attempt to reduce the SLs at the source or along the path from the source to receiver. The worker is exposed for the same length of time, but since the SL is reduced, less noise exposure occurs. Engineering controls might include quieter motors, reduced impact noise, barriers, sound-absorbing material, control rooms, and reduced metal-to-metal contact. Administrative controls attempt to remove the worker from the noisy area. In this case, the SLs remain the same, but since the worker does not spend as much time near the noise, the overall noise exposure is reduced. Administrative controls may include minimizing or eliminating tasks in high SL areas, remote monitoring, training to allow for faster task completion, and job rotation. In most cases, the most effective noise reduction is the use of an engineering control that reduces the SL. When all feasible engineering controls are implemented, administrative noise controls are the next best choice. One starting point for identifying specific engineering and administrative controls is MSHA Program Information Bulletin No. P08–12, "Technologically Achievable, Administratively Achievable, and Promising Noise Controls (30 CFR Part 62)" [Stricklin et al. 2008].

Intervention Examples

One of the workers in underground coal mining likely to experience noise overexposures is the continuous mining machine operator. Exposure data collected by NIOSH showed that continuous mining machine operators had MSHA PEL doses as high as 347% [Bauer et al. 2006]. Some of the interventions that have been implemented include coated flight bars and coated tail rollers (engineering controls) and remote operator positioning and job rotation (administrative controls).

In another case, plant operators in coal preparation plants had MSHA PEL doses as high as 221% [Bauer et al. 2006]. The plant operators are the workers that routinely travel throughout the plant checking processes, cleaning, and performing minor maintenance. They are exposed to many different SLs emanating from any number of different pieces of processing equipment. Some of the interventions include: urethane screen decking, chute liners, and enclosing noise sources such as air compressors and pumps (engineering controls); control booths, remote monitoring, restricting travel to lowest SL areas, and job rotation with control room operators (administrative controls). Finally, in surface coal mining, dragline oilers had MSHA PEL measured doses as high as 193% [Bauer et al. 2006]. Interventions have included: sound barrier curtains around motor-generator sets, quieter motor blower fans, and wrapping of ductwork with insulation (engineering controls); restricting oiler time in the "house," remote greasing, and job rotation with dragline operator and dozer operator (administrative controls).

Intervention Success

Judging intervention success is accomplished by conducting a time study after a noise control has been implemented, then comparing the postintervention time study with the preintervention time study. For inst7ance, if an engineering control has been implemented, one would look to see if the accumulated dose has been reduced for the task that is completed where the equipment has had the engineering control applied. If an administrative control has been employed as the exposure reduction method, one would simply check to see if the time spent at that location or task is less and, subsequently, if the result is less dose accumulation. DOSES allows these types of analyses to be completed rather easily by comparing pre- and postintervention tables or dose graphs.

USER INSTRUCTIONS

1. Start DOSES (Splash Screen)

When the DOSES program is opened it displays the **Start DOSES** splash screen that shows three button choices: **OPEN Time Study, NEW Time Study,** and **Continue.**

To open a saved file, select the **OPEN Time Study** button. A file browse dialog is displayed to allow selection of a previously saved study (go to 2).

To enter a new study, select the **NEW Time Study** button. A properties dialog is displayed for defining a new study (go to 3).

To close this screen, select the **Continue** button, the **Start DOSES** screen closes, and the **DOSES Time Study Editor** screen is displayed in a disabled mode. From here select **Open, New** or **Exit** from the **File** menu, or the **Help** files can be viewed. The other menus will be disabled until an existing study is opened or a new study is created.

Start DOSES Screen

When the DOSES program is selected, the program opens and displays a **Start DOSES** splash screen that shows three button choices: **OPEN Time Study, NEW Time Study,** and **Continue.** Clicking in the dialog box "**Do not show this dialog in the future**" will prevent this screen from coming up when DOSES is opened again.

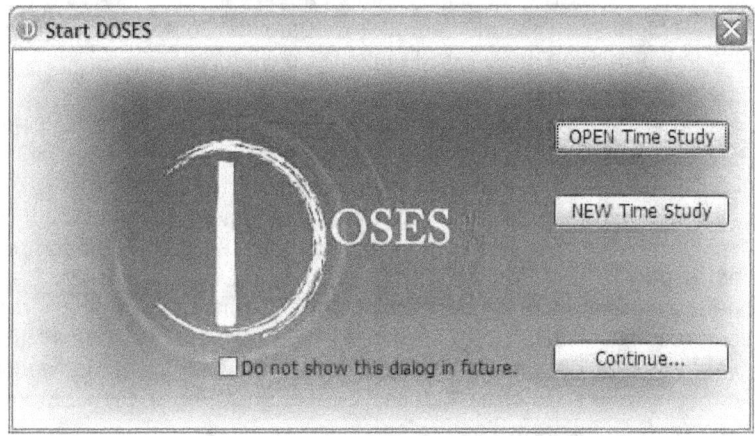

2. OPEN Time Study

After selecting **OPEN Time Study**, the **Select Study to Open** dialog opens.
A saved study can be opened by clicking on the study file and clicking the **OK** button or double-clicking on the study file (go to 4).

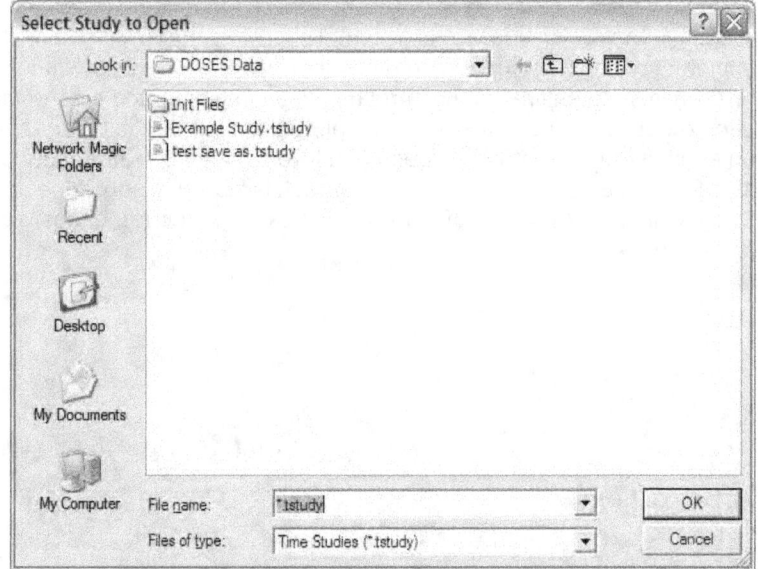

3. Time Study Properties - New Study Screen: Time Study Tab

After selecting **NEW Time Study**, the **Time Study Properties - New Study** screen opens on the first time study property tab: **Time Study**. This window is the default screen when **NEW Time Study** is selected from the **Start DOSES** screen. The other tabs—**Employee, Mine, Shift,** and **Dosimeter**—contain additional information to complete about the time study. After all of the desired information is entered, clicking the **OK** button moves the user to the **DOSES Time Study Editor** screen to enter the time-motion data.

When the **Time Study Properties - New Study** screen is opened, the current date is automatically inserted by the program and shown on the screen for viewing and documentation purposes. The **Creation Date** cannot be changed by the user.

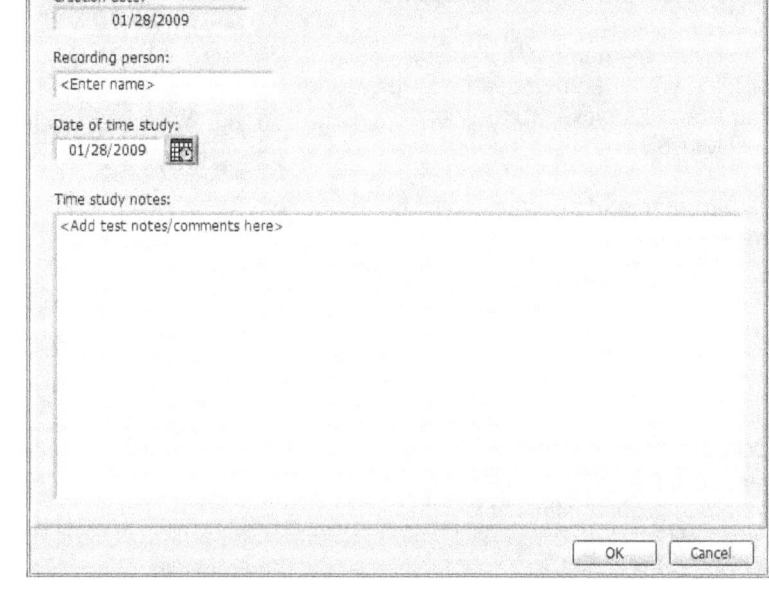

The name of the person doing the study is entered under the **Recording Person** window.

The date of the study is entered under the **Date of Time Study** window. This can be done manually typing in the date or by clicking on the calendar icon to the right of the dialog box and selecting the date.

Notes or comments that were taken during the actual time-motion study can be added now or later in the text box under **Time Study Notes**.

Time Study Properties - New Study Screen: Employee

The next time study property tab to be selected and filled out is **Employee**. The Employee property contains five parameters: **Employee Name, Employee ID, Work Location/Section, Occupation,** and **Equipment Operated**.

The employee name and ID being studied are entered under the **Employee Name** and **Employee ID** text box, respectively.

As part of the time-motion study, the work location is recorded in the **Work Location/Section** text box.

The employee's official MSHA occupation/title and equipment operated code is selected from the **Occupation** list box and the **Equipment Operated** list box by clicking on the appropriate line.

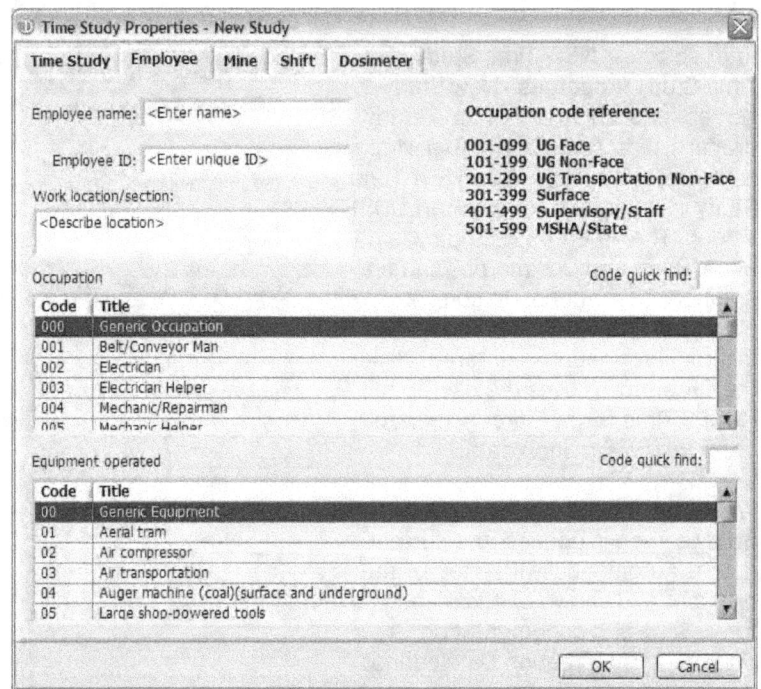

A **Code Quick Find** text boxes located within the **Employee** property are available to quickly set the **Occupation** and **Equipment Operated** codes. The **Code Quick Find** text box for the Occupation is only available when entering a new study. The Occupation code cannot be changed once the properties dialog has been closed when creating a new time study.

Time Study Properties - New Study Screen: Mine Tab.

The Mine property contains three parameters: **Company Name, Mine Name,** and **Mine MSHA ID Number**. The information can be entered into the text boxes immediately or at a later time.

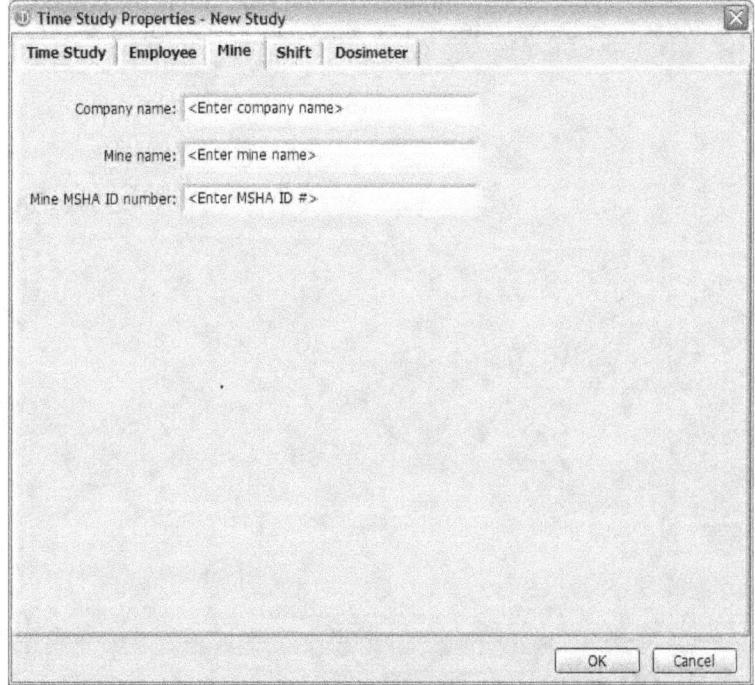

8

Time Study Properties - New Study Screen: Shift

The Shift property contains four parameters: **Shift Start**, **Shift Length**, **Shift Tonnage (Raw)**, and **Shift Tonnage (Clean)**.

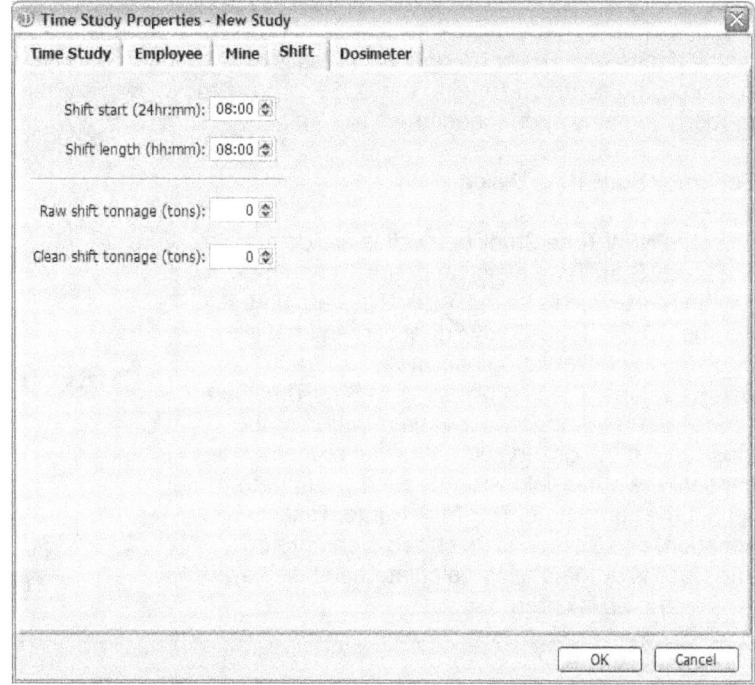

Time Study Properties - New Study Screen: Dosimeter Tab

If a dosimeter was not used, click **OK** at the bottom of the screen (go to 4).

The Dosimeter property tab specifies the dosimeter characteristics. These include the Dosimeter **Manufacturer, Model, Serial #, Log File** information, and **Calibration and Exposure Details**. To make changes, select the desired parameter and make the change using the keyboard or mouse.

Enter the dosimeter Manufacturer, Model and Serial # in the test boxes under **Dosimeter.**

Enter the dosimeter's pre- and post-calibrations, dose results and run time information in the appropriate text boxes under **Calibration and Exposure Details.**

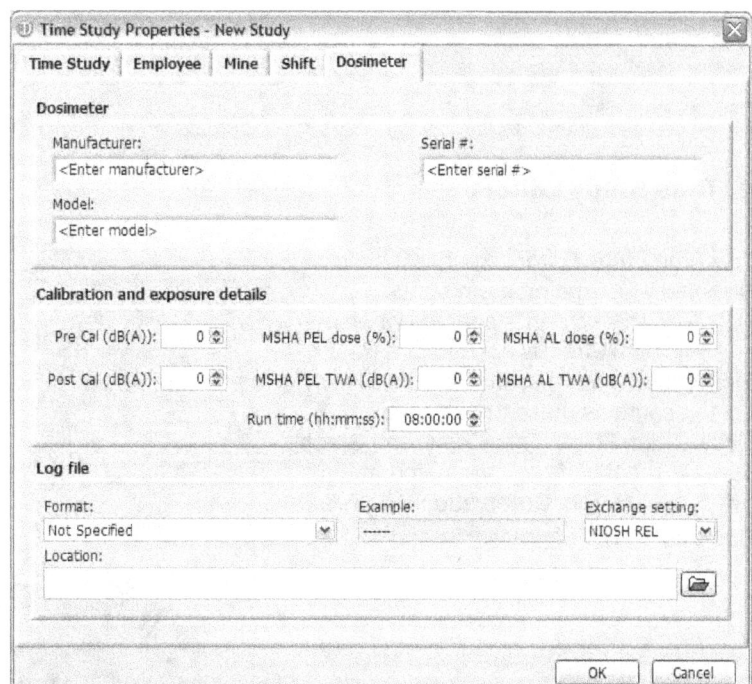

> *NOTE:* The calibration information is used for record keeping only, and is not factored in to the program calculations.

If a log file was extracted from the dosimeter for comparison use in the DOSES program, complete the information under **Log File**. Pay close attention to specifying the format of the ASCII file exported from the dosimeter or the data may not import in to DOSES correctly.

NOTE: The dosimeter log file information is not needed to complete the analysis.

If changes were made mistakenly, press the **Cancel** button. When the **Cancel** button is pressed a confirmation dialog opens with **Yes** or **No** options. If **Yes** is chosen, the properties dialog is closed without saving the changes and control returns to the DOSES Time Study Editor screen. If **No** is chosen, then the Time Study property remains open for editing.

Set Study Start Time Dialog

After the **NEW Time Study** properties are defined, a **Set Study Start Time** dialog appears and prompts for the **Start Time** to be set. After the **Start Time** is set the **DOSES Time Study Editor** screen enables and sets the observation parameters to their default values. The first observation **Stop Time** can now be set by either entering the duration or the stop time of the observation. Whichever method is chosen, the other field will automatically update to reflect the same time. The **Task Name**, **Task Location**, and **SL** (sound level) are then entered. The Dose is automatically calculated and displayed in **%** as the **SL** is entered.

4. Time Study Editor

The **Time Study Editor** screen is initially disabled when the program starts. However, this screen is enabled when an existing **Time Study** is opened or a **New Time Study** is successfully started. This screen contains these three areas: **Observation Editor**, **Table View**, and **Graph View**. At the top of the screen, four menus are available: **File**, **Observation**, **Analysis**, and **Help**.

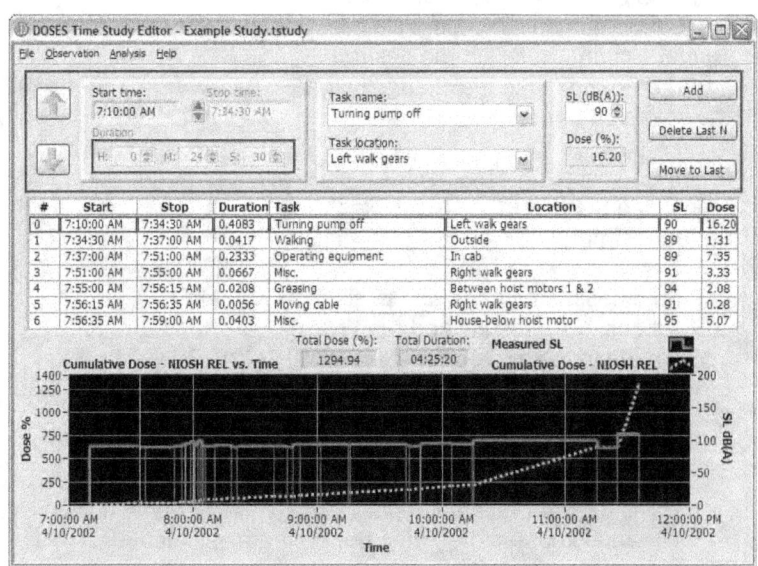

Time Study Editor: File Menu

The **File** menu contains the following options:

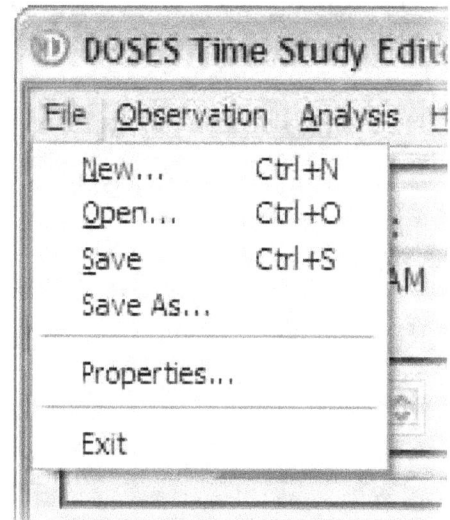

Option	Description
New:	Creates a new time study. This will automatically bring up the Time Study Properties dialog.
Open:	Opens an existing time study using the File Browse dialog.
Save:	Saves the current time study.
Save As:	Saves the current time study under a different name. A File Save dialog is opened, which permits the name of the file to be changed.
Properties:	Activates the Time Study Properties dialog to allow the user to view or modify the properties of the current time study.
Exit:	Exits the DOSES program.

Time Study Editor: Observation Menu

The **Observation** menu contains the following options:

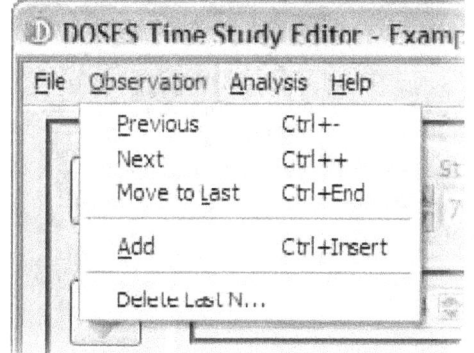

Option	Description
Previous:	Moves to the previous observation in the list.
Next:	Moves to the next observation in the list. If already at the last observation, this option will create a new observation entry and add it to the end of the list.
Move to Last:	Moves directly to the last observation in the list.
Add:	Adds a new observation to the end of the list.
Delete Last N:	Deletes a number (N) of observations from the list using the Delete Last N dialog.

Time Study Editor: Analysis Menu

The **Analysis** menu contains the following options:

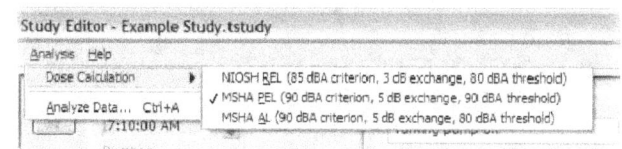

Option	Description
Dose Calculation:	Provides an expanded list to choose the active dose calculation. The options are:
NIOSH REL	Uses an 85-dB(A) criterion level with a 3-dB tradeoff.
MSHA PEL	Uses a 90-dB(A) criterion level with a 5-dB tradeoff.
MSHA AL	Uses a 90-dB(A) criterion level with a 5-dB tradeoff.
Analyze Data	Activates the **Time Study Analysis** dialog.

NOTE: *The active dose calculation can be changed at any time.*

Time Study Editor: Help Menu

The **Help** menu contains the following options:

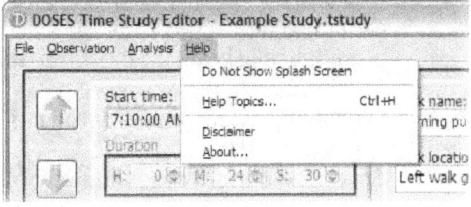

Option	Description
Do Not Show Splash Screen:	Toggles On/Off the option to show the Start DOSES Screen on program start.
Help Topics:	Displays the online help.
Disclaimer:	Displays the NIOSH Disclaimer dialog.
About:	Displays the About DOSES dialog.

Observation Editor Area

This area is initially disabled when the program starts. The screen is enabled when an existing Time Study is opened or a New Time Study is successfully started.

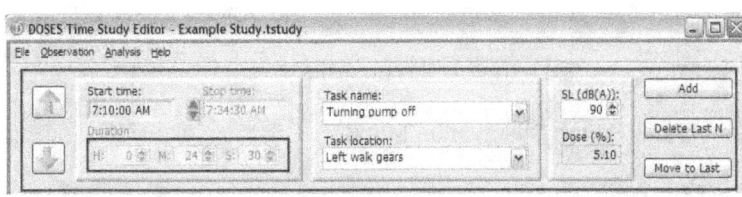

The **Observation Editor** contains five main areas:

1. Previous/Next Observation
2. Start/Stop Time & Duration
3. Task Name & Task Location
4. SL & Dose
5. Editing Buttons

Previous/Next Observation

Pressing the up or down yellow arrows allow for movement backward and forward in the list of observations.

Start/Stop Time & Duration Text Boxes

Each observation is defined in terms of a **Start Time** and **Stop Time**. The **Start Time** is set once, when initializing the first observation with the **Set Study Start Time** dialog. From then on, the observation **Start Time** is automatically set based on the **Stop Time** of the previous observation. The **Stop Time** is set as part of defining a new observation. It can be set by entering the exact stop time or by entering the **Duration** in hours, minutes, and seconds. Regardless of how the **Stop Time** is set, the other stop time fields will update automatically.

> NOTE: Only the **Stop Time** of the last observation can be changed. If the **Stop Time** of an observation within the study needs to be changed, all observations following must be deleted and reentered. This is to ensure time continuity in the observations.

Task Name & Task Location List Boxes

The task name and task location are entered for each observation. Entry is made by selecting the list box drop-down arrow and choosing from the list.

A short explanation is displayed inside the list box for quick assistance. When starting a NEW Time Study with a previously unused occupation code, the "**Add/Edit...**" option is the only option available for both **Task Name** and **Task Location**. After selecting **Yes** to the **Do you want to Edit Task List?** dialog, an **Edit Item List** dialog appears, allowing for items to be added or removed from the list. Once the **Add/Edit** is completed, the new item(s) can be selected throughout the time study. The **Task Name** and **Task Location** list boxes are initially empty when creating a study using a new employee occupation code. They are populated as the observations are entered and automatically saved for future use according to the employee occupation code.

> *NOTE:* Removing an item from the list box does not remove it from the observations in the time study. Those entries will have to be manually reset to a valid list box value.

Sound Level and Dose

Sound level (SL) is a way of measuring the sound pressure (and perceived loudness of sounds) on a logarithmic scale, usually in dB units. However, in this case, dB(A) is used to measure the sound. This represents the A-weighting of the SL.

Each observation contains an SL dB(A) number that is entered by either typing the number in the area provided or by using the numeric up/down control and moving to the desired number. The numeric up/down control increases or decreases the number by 1 dB(A).

Dose represents the **%** of noise exposure based on the duration and SL of the observation. It is calculated based on one of three formulas chosen from the Analysis menu:

$$\text{NIOSH REL dose} = 0 \text{ for dB(A)} < 80$$
$$= (\text{Duration} / (8 / (2^{(L_{eq} - 85)/3}))) * 100$$

$$\text{MSHA PEL dose} = 0 \text{ for dB(A)} < 90$$
$$= (\text{Duration} / (8 / (2^{(L_{eq} - 90)/5}))) * 100$$

$$\text{MSHA AL dose} = 0 \text{ for dB(A)} < 80$$
$$= (\text{Duration} / (8 / (2^{(L_{eq} - 90)/5}))) * 100$$

where duration is in units of hours and L_{eq} is the SL taken from a sound level meter or dosimeter data.

> *NOTE:* The dose is calculated by the program from the SL and duration of the exposure.

Editing Buttons

There are three Editing Buttons: **Add, Delete Last N**, and **Move to Last**. They are located on the top right of the DOSES Time Study Editor screen.

The **Add** button automatically adds a new observation to the end of the list with the Start Time set to the Stop Time of the previous observation and the duration set to 1 minute. It is then possible to continue to set **Stop Time, Task Name, Task Location**, and **SL** for that observation.

The **Delete Last N** button, when pressed, brings up the Delete Last N Observations dialog where the number of observations to be deleted (from the end of the study) is specified. This number can be entered by typing in the preferred number or by clicking on the numeric up/down control.

The **Move to Last** button sets the current observation to the last one entered into the time study.

Table View

The **Time Study Table View** is located in the center of the **DOSES Time Study Editor** screen. The **Table View** provides a tabular view of the observations that were entered into the time study. Movement around the **Table View** is possible. The table is updated automatically as observation data are entered. New data cannot be entered by clicking on the last row of the table. The current observation will have a blue border around it.

Graph View

The **Time Study Graph View** is located at the bottom of the **DOSES Time Study Editor** screen. The graph provides a plot of the observed **SL** and cumulative dose over the duration of the time study. The graph is updated automatically as observation data are entered.
The graph crosshair will be centered on the **Stop Time** of the current observation and anchored to the cumulative dose plot.

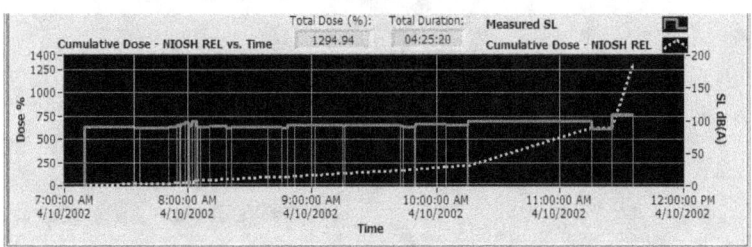

> *NOTE:* Positioning the mouse cursor over the blue vertical or horizontal line of the crosshair and holding the left button down on the mouse will allow the user to move the crosshair to access the graph. In the **Time Study Table** the row chosen by the graph crosshair will have a blue border around it.

After all the data are entered, the **Analysis** menu can be selected. From the **Analysis** drop-down menu, left-clicking on **Analyze Data** will activate the **Time Study Analysis** dialog.

Time Study Analysis Dialog

The Analysis dialog contains six tabs: **Summary, Raw Data, Study vs. Dosimeter, Exposure Comparison, Study vs. Study,** and **Printing & Export**. Selecting **Analyze Data** from the Analysis menu on the **DOSES Time Study Editor** screen permits access to this dialog. The **Summary** tab is displayed by default when **DOSES Time Study Analysis** opens. The file name is displayed in the title bar of the dialog for reference.

5. Time Study Analysis Dialog – Summary Tab

In the Summary section, the analyzed data are shown in two separate tables—one where the breakdown is by **Task**, the other where the breakdown is by **Location**. Included for reference beside the **Breakdown By Task** table is the **Overall Noise Exposure** (%) generated from the time study.

The **Breakdown by Task** table has a row for each unique **Task Name** in the time study. Columns associated with each row include the **# Obs.**, **Total Time**, **% Dose**, and **L**$_{AVG}$ **dB(A)**.

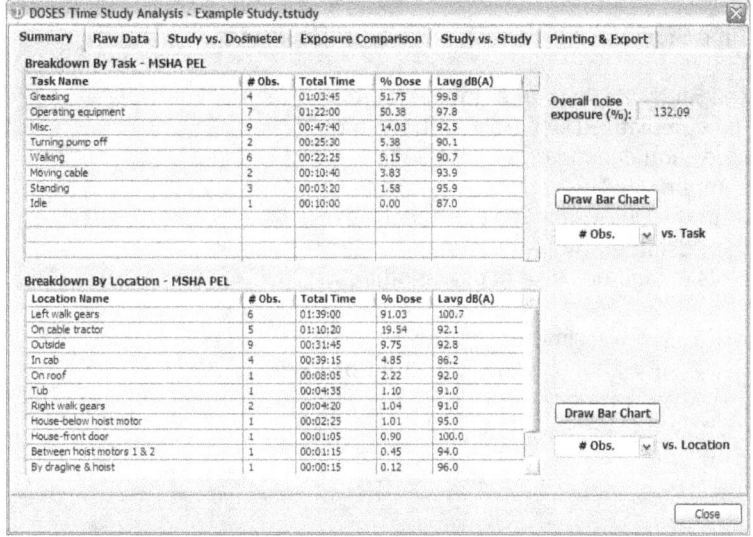

 # Obs.: The total number of observations where the same task was performed during the time study.
Total Time: The total time where the same task was performed during the time study.
% Dose: The total dose received by the subject during a specific task during the time study.
L$_{AVG}$ **dB(A):** The average of the A-weighted SLs where the same task was performed during the time study.

The table can be sorted by clicking on any of the column headers. Clicking the same column header multiple times will toggle the sort between ascending and descending order.

It is possible to visually chart the information contained in the table by clicking the **Draw Chart** button located below the table. The adjacent list box provides the option to chart the **# Obs., Total Time, % Dose**, or **L**$_{AVG}$ **dB(A)** versus the **Task Name**. To change the category, select from the list box by using the drop-down arrow.

Time Study Analysis Dialog – Raw Data Tab

The **Raw Data** section is a tabular view of the time study data. It is a larger version of the **Table View** found on the **DOSES Time Study Editor** screen. In the **Raw Data** section, the table can be sorted by clicking on any of the column headers. Clicking the same column header multiple times will toggle the sort between ascending and descending order.

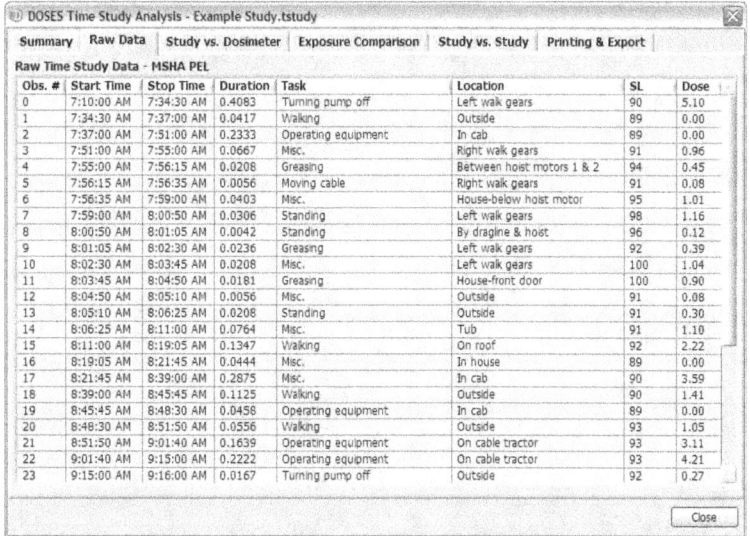

Time Study Analysis Dialog – Study vs. Dosimeter

The **Study vs. Dosimeter** tab is where the cumulative **Dose** from both the time study and dosimeter are plotted for comparison. The dosimeter data comes from the **Dosimeter Log File** specified in the **Dosimeter** property. Ideally, the two plots should be close to one another.

NOTE: If no Dosimeter Log File is specified, only the time study plot will be shown.

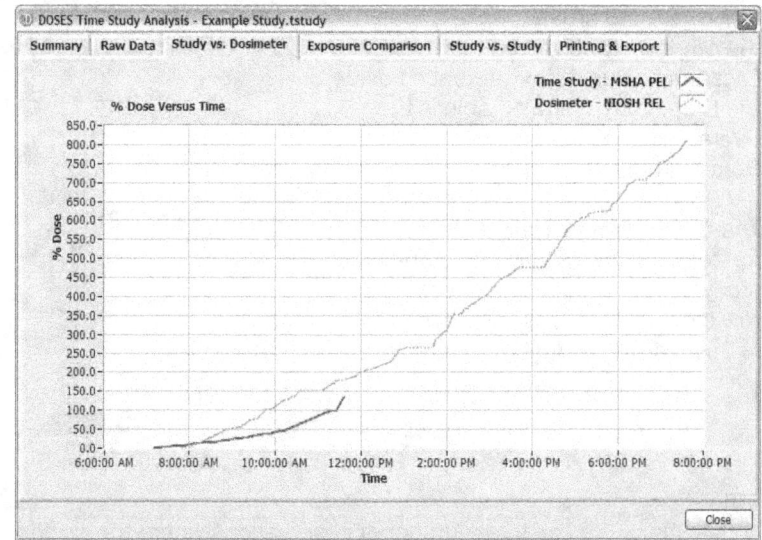

Time Study Analysis Dialog – Exposure Comparison

The **Exposure Comparison** tab displays a plot of the three different calculations (NIOSH REL, MSHA PEL, and MSHA AL) using the time study data. The purpose is to provide a comparative view of the three methods using the same data.

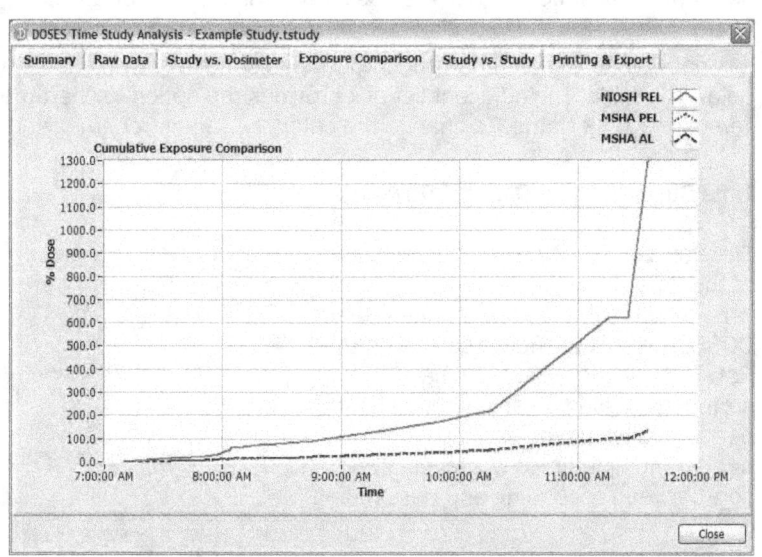

Time Study Analysis Dialog – Study vs. Study

The **Study vs. Study** tab enables a comparison of the current time study with another. The comparison is on the **Breakdown by Task**. Selection of **Study 2** is done by clicking on the yellow folder in the middle of the screen. The tables can be sorted in ascending or descending order by clicking on the column headings. The **Overall Noise Exposure** section is shown for both studies.

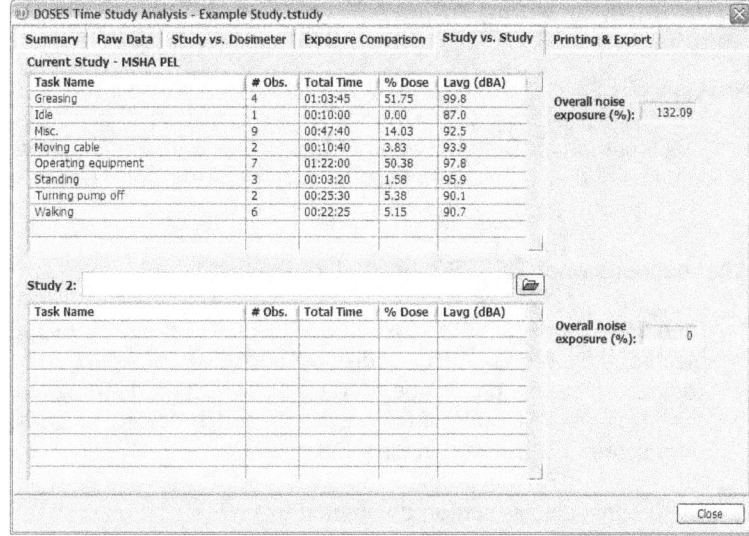

Time Study Analysis Dialog – Printing & Export

The **Printing & Export** tab has the controls for printing reports and exporting the data.

Printing a Report

There are two report options for printing: **Study Report** and **Study vs. Study**. Specific options can be chosen for what parts to include in the study report by clicking the appropriate checkboxes.

After the report option is chosen, left-click the **Print** button to access the print dialog. From this dialog, select whether to print to a file (HTML) or selected printer and the page orientation.

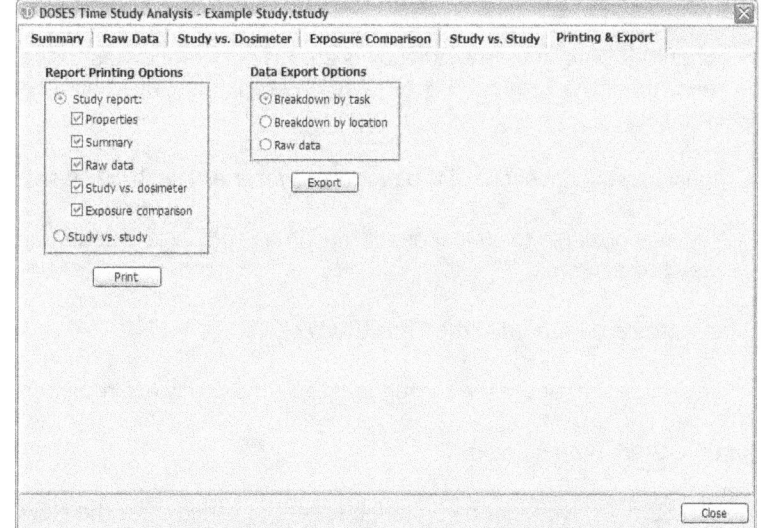

Exporting Time Study Data

There are three options for which data to export:

- Breakdown by **Task** results from the **Summary** tab
- Breakdown by **Location** results from the **Summary** tab
- Raw data as seen in the table on the **Raw Data** tab

The data are exported in an ASCII comma-delimited format with a default .csv extension which makes it readable by Microsoft Excel. After selecting the export option, left-click the **Export** button to access the export file dialog to select or choose to destination file.

6. Frequently Asked Questions

What is DOSES?

DOSES stands for Determination of Sound Exposures. It is a program designed to analyze tasks and locations of task sound exposures associated with them. Based on these data, the program can calculate worker sound exposure during a miners' work shift.

What happens when DOSES is opened for the first time?

When DOSES opens for the first time, a display Start DOSES splash screen appears. If OPEN Time Study is selected, a file browse dialog is displayed to allow for selection of a previously saved study. When NEW Time Study is selected, a properties dialog is displayed for defining a new study. When opened, the Time Study property is displayed first. If Cancel is selected, the Start Doses screen closes and the DOSES Time Study Editor screen is displayed in disabled mode.

Can the original display screen be opened again?

No, the original display screen or splash screen is opened only when the DOSES icon is selected from the desktop. However, if a new Time Study is desired, open the File menu and select New.

Why is the Date of Time Study grayed out?

When the Date of the Time Study is grayed out, it indicates that the user is not in a NEW Time Study. Once the Date of the Time Study has been entered, it is grayed out on the screen to indicate that it cannot be changed.

Is it possible to cancel out of the properties after a New Study is selected?

Yes, it is possible to cancel out of the Time Study properties at any time. DOSES will then open to the previous screen.

What notes are entered in the Time Study?

Any notes that are taken throughout the Time Study are recorded here.

Can the Start Time be edited?

The Start Time cannot be edited; it is set only once after the New Time Study properties have been determined. From then on, the observation Start Time is automatically set based on the Stop Time of the previous observation.

Can the Stop Time be edited?

Yes, the Stop time can be edited. After the initial Start Time is set, a Stop Time or duration must be entered for each task. However, since the Start Time is determined by subsequent tasks, only the Stop Time of the last observation can be changed. If the Stop Time of an observation within the study needs to be changed, all of the subsequent observations must be deleted and reentered.

Why are some of the boxes on the DOSES Time Study Editor screen grayed out?

When a NEW Time Study is active and information is being entered, there are five areas that can be added or changed: Stop Time or duration, Task Name, Task Location, and SL. The Start Time and Dose are grayed out, but remain visible for viewing purposes. However, if viewing a previously saved time study, only the

Task Name, Task Location, and SL can be edited. Therefore, the other boxes are grayed out, indicating that changes cannot be made.

Can a previously saved Time Study be opened?

Yes, a previously saved Time Study can be opened. Time Studies are saved in a file with a .tstudy extension.

What is Code Quick Find?

Code Quick Find is a text box located in the Employee property within the Occupation and the Equipment Operated area. Typing in the number quickly displays the desired area of selection.

Why is the Occupation Code Quick Find text box grayed out?

When the Code Quick Find area is grayed out, it indicates that the mandatory Occupation selection was made and therefore cannot be changed.

Can previously saved Task Names or Task Locations be reused?

Yes, once any Task Name or Task Location is entered, it is available to be used again within the same study.

What do the letters "SL" refer to?

"SL" stands for sound level. A sound pressure level taken from an SLM or dosimeter that has been processed through the instrument's A-weighting filter is referred to as a "sound level" (SL).

What is TWA?

TWA stands for time-weighted average. The 8-hr time-weighted average sound level, dB(A), can be calculated from the daily dose (D) according to the following formula: TWA (8) = 10 × Log (D/100) + 85 for the NIOSH REL. The formula for the MSHA REL is: TWA (8) = 16.61 × Log_{10} (D/100) + 90. The TWA (8) dB(A) is used so that all work shifts regardless of length can be easily compared.

What does dB(A) refer to?

dB(A) refers to decibels measured with A-weighting. dB(A) is a commonly used frequency weighting that closely approximates the frequency response of the human ear. The human ear does not perceive sounds of equal sound pressure level as being equally loud if the frequencies are different. The ear is less sensitive at low frequencies.

A-weighting is commonly used to measure environmental and industrial noise when assessing potential hearing damage and other noise effects at moderate- to high-intensity levels.

Before printing a chart, can the color be changed?

Yes, the color can be changed by selecting the Draw Bar Chart located on the DOSES Time Study Analysis screen. Click on the colored text box labeled Bar color and select the desired color for the chart. This is located in the upper right corner of the DOSES Chart Breakdown Data dialog.

Why does the Observation # begin at 0?

The initial Start Time is set and is defined as observation number 0. Therefore, each observation begins at 0, representing the first entry.

What are the yellow up/down arrows used for?

After the NEW Time Study properties are defined, a Set Study Start Time dialog appears and prompts for the Start Time to be set. After the Start Time is set, the yellow up/down arrows are used to enter all subsequent observations.

NOTE: The first Observation # always begins at 0.

What does the Delete Last N button do?

The Delete Last N button, located on the Observation Editor screen when selected, will bring up the Delete Last N Observations dialog where the number of observations to delete is set. Upon confirmation, the observations are deleted.

NOTE: Observations can only be deleted from the end of the time study due to the time dependence of the start and stop times.

REFERENCES

64 Fed. Reg. 49548 [1999]. Mine Safety and Health Administration: health standards for occupational noise exposure; final rule (30 CFR Parts 56, 57, 62, 70, and 71).

Bauer ER, Babich DR, Vipperman JS [2006]. Equipment noise and worker exposure in the coal mining industry. Pittsburgh, PA: U.S. Department of Health and Human Services, Centers for Disease Control and Prevention, National Institute for Occupational Safety and Health, DHHS (NIOSH) Publication No. 2007–105, IC 9492.

MSHA [2001]. Information resource center. Part 50: diskette user's handbook. Available at: http://www.cdc.gov/niosh/mining/data/pdfs/codes.pdf

NIOSH [1998]. Criteria for a recommended standard: occupational noise exposure – revised criteria 1998. Cincinnati, OH: U.S. Department of Health and Human Services, Centers for Disease Control and Prevention, National Institute for Occupational Safety and Health, DHHS (NIOSH) Publication No. 98–126

Stricklin KG, Quintana FA, Skiles ME [2008]. Technologically achievable, administratively achievable, and promising noise controls (30 CFR part 62). Arlington, VA: Mine Safety and Health Administration, Program Information Bulletin No. P08–12, June 18, 2008. Available at: http://www.msha.gov/regs/complian/PIB/2008/pib08-12.pdf

APPENDIX.—MSHA EQUIPMENT AND OCCUPATION CODES

MSHA Equipment Codes [MSHA 2001]

Code	Equipment
00	Generic equipment
01	Aerial tram
02	Air compressor
03	Air transportation
04	Auger machine (coal) (surface and underground)
05	Large shop-powered tools
06	Blow pipe/gun (air blasting)
07	Boats, barges
08	Bulldozer
09	Carriage-mounted drills
10	Chute
11	Classifier, cyclones
12	Continuous miner, tunnel borer
13	Conveyor
14	Crane, derrick
15	Crusher, breaker, ball and rod mills
16	Cutting machine
17	Dimension stone cutting/polishing machinery
18	Dredge
19	Elevator, skip, cage
20	Electric/hydraulic coal drills
21	Fan
22	Flotation and filters
23	Forklift
24	Front-end loader
25	Gathering arm loader
26	Grizzlies
27	Gunite, shotcrete machine
28	Handtools (not powered)
29	Handtools (powered)
30	Hoist car, dropper
31	Hydraulic jets (monitors)
32	Impactor (pneumatic)
33	Load-haul-dump, scoop, bobcat
34	Locomotive
35	Longwall machine (plow, shearer, shield, stageloader)
36	Longwall subparts (jacks, chocks)
37	Mantrip, jeep
38	Man lift (not elevator)
39	Mill grinding
40	Milling machine
41	Mine car (supply)
42	Mine car (ore, coal)
43	Mucking machine
44	Ore haulage trucks (off highway, underground)
45	Ore haulage trucks (highway)
46	Packaging machine
47	Pneumatic blasting agent loader
48	Pump
49	Raise borer
50	Raise climber
51	Raw coal storage
52	Road grader
53	Rock drill (jackleg)
54	Roof bolting machine
55	Rock dusting machine
56	Rotary dump
57	Scraper, pan
58	Screen
59	Shortwall machine
60	Shovel (stripping, dragline, bucket-wheel)
61	Shuttle car
62	Skip pocket
63	Slusher, scraper, scram
64	Tamping machine, railroad
65	Track maintenance and repair
66	Tractor (underground)
67	Trucks, all other
68	Tugger air winch
69	Washers
70	Welding machine
71	Machine, NEC

MSHA Occupation Codes [MSHA 2001]

Code	Occupation	Code	Occupation
000	Generic occupation	057	Stope Miner
001	Belt/Conveyor Man	058	Drift Miner
002	Electrician	059	Raise Miner
003	Electrician Helper	063	Miner, NEC
004	Mechanic/Repairman	064	Contract Miner
005	Mechanic Helper	067	Power Shovel Operator
006	Rock Duster	068	Bulldozer Operator
007	Blaster/Shooter/Shotfirer	070	Auger Operator
008	Stopping Builder/Ventilation Man/Mason	071	Auger Helper
009	Supplyman	072	Mobile Bridge Operator
010	Timberman/Jacksetter/Propman	074	Tractor Operator/Motorman
011	Wireman	075	Grader operator/Roadgrader operator
015	Fan Attendant	076	Truck Driver
016	Laborer	078	Crane Opr./Dragline/Backhoe
019	Cement Man	082	Frontend Loader Operator
024	Trainee	101	Belt/Conveyor Man
025	Bobcat Operator	102	Electrician
026	Grizzly Man/Car Dump Opr	103	Electrician Helper
028	Scoop Tram-Load Haul Opr	104	Mechanic/Repairman
029	Mucking Machine Opr	105	Mechanic Helper
030	Slusher Operator	106	Rock Duster
031	Shotfirer Helper	108	Stopping Builder/Ventilation Man/Mason
032	Brattice Man	109	Supplyman
033	Coal Drill Helper	110	Timberman/Propman/Jacksetter
034	Coal Drill Operator	111	Wireman
035	Continuous Miner Helper	112	Belt Vulcanizer
036	Continuous Miner Operator	113	Cleanup Man
037	Cutting Machine Helper	114	Coal-MN Sampler
038	Cutting Machine Operator	115	Fan Attendant
039	Hand Loader	116	Laborer
040	Headgate Operator	117	Rodman
041	Jack Setter (Longwall)	118	Oiler/Greaser
042	Loading Machine Helper	119	Welder
043	Loading Machine Operator	122	Coal Dump Operator
044	Longwall Shearer Operator	123	Transit Man
045	Rockman	124	Trainee
046	Roof Bolter/Rock Bolter	126	Grizzly Man/Car Dump Opr
047	Roof Bolter Helper	128	Load Haul Dump Opr/Gizmo
048	Roof Bolter Mounted	149	Bullgang Foreman/Labor Foreman
049	Section Foreman	154	Belt Cleaner
050	Shuttle Car Operator	155	Chainman
051	Stall Driver	156	Rock Driller
052	Tailgate Operator	157	Pumper
053	Utility Man	158	Rock Machine Operator
054	Scoop Car Operator	159	Water Line Man

160	Shopman/Machinist		326	Forklift Operator
163	Miner NEC		327	Surface Miner
167	Power Shovel Operator		328	Utility Man
168	Bulldozer Operator		329	Vacuum Filter Operator
175	Grader Operator		330	Skip Tender
176	Truck Driver		331	Clam/Claw Operator
178	Crane Opr/Dragline/Backhoe		333	Drill Helper
182	Frontend Loader Operator		334	Drill Operator
201	Belt/Conveyor Man		340	Boom Operator
216	Trackman		341	Belt Man/Conveyor Man
220	Cager		342	Bit Sharpener
221	Hoistman/Engineer		343	Car Trimmer/Car Loader
224	Transportation Trainee		344	Car Shake-Out Operator
230	Skip Tender		345	Crusher Attendant
240	Loader Head/Roscoe Operator		347	Froth Cell Operator
255	Chainman		348	Machinist
261	Battery Station Operator		349	Rotary Dump Operator
262	Brakeman/Roperider/Snapper		350	Shuttle Car Operator
263	Track Foreman		351	Scoop Operator
265	Dispatcher		352	Steel Worker
269	Motorman/Swamper/Switchman		354	Sweeper Operator
276	Driver/Tractor Opr/Jeep		355	Chainman
277	Buggy Pusher		356	Rock Driller
301	Conveyor Operator		357	Washer Operator
302	Electrician		358	Water Circuit Operator
303	Electrician Helper		359	A Self-Propelled Compactor Operator
304	Mechanic/Repairman		360	Machinist
305	Mechanic Helper		362	Brakeman/Trip Rider
306	Welder (Non-Shop)		363	Miner NEC/Quarry Worker
307	Blaster/Shooter/Shotfirer		365	Dispatcher
308	Mason		366	Waterboy
309	Supplyman/Supply Truck Driver/Warehouseman		367	Power Shovel Operator
			368	Bulldozer Operator
310	Scraper Operator		369	Motorman/Locomotive Operator
311	Wireman		370	Auger Operator
312	Belt Vulcanizer		371	Auger Helper
313	Cleanup Man		372	Barge Attendant/Boat/Dredge
314	Coal Sampler		373	Car Dropper
315	Fan Attendant		374	Cleaning Plant Operator
316	Laborer/Utility Man/Pumper		375	Road Grader Operator
317	Rodman		376	Truck Driver
318	Oiler/Greaser		378	Washer Operator
319	Welder (Shop)		379	Dryer Operator
320	Cage Attendant/Cager		380	Fine Coal Plant Operator
321	Hoistman/Engineer		381	Hoist Operator Helper
322	Dump Operator		382	High Lift Operator/Front End Loader
323	Transit Man		383	Highwall Drill Helper
324	Trainee		384	Highwall Drill Operator
325	Bobcat Operator		385	Lampman

386	Refuse Truck Driver/Backfill Truck Driver	430	Assistant Mine Foreman/Asst Mine Mngr
387	Rotary Bucket Excavator Operator	449	Mine Foreman/Mine Manager
388	Scalper-Screen Operator	456	Engineer (Electricity/Ventilation/Mining)
389	Forklift Operator	462	Fire Boss Pre-Shift Examiner
390	Silo Operator	464	Inspector
391	Stripping Shovel Operator	481	Superintendent
392	Tipple Operator	489	Outside Foreman
393	Weighman	494	Preparation Plant Foreman Safety Director
394	Carpenter	495	Safety Director
395	Water Truck Operator	496	Union Representative
396	Watchman/Guard	497	Outside Foreman
397	Yard-Engine Operator	590	Education Specialist
398	Dimnsn Stone Cutter/Polisher	591	Mineral Industrial Safety Officer
402	Master Electrician	592	Mine Safety Inspector
404	Master Mechanic	593	Safety Representative
414	Dust Sampler/Lab Technician	594	Training Specialist
418	Maintenance Foreman	999	Unknown or NEC
423	Surveyor		